UPDOG

WORLD'S GROSSEST

WORLD'S GROSSEST CREEPY CRAWLERS

Scott Nickel

Lerner Publications ◆ Minneapolis

Lerner Publications Company
An imprint of Lerner Publishing Group, Inc.
241 First Avenue North
Minneapolis, MN 55401 USA

For reading levels and more information, look up this title at www.lernerbooks.com.

Main body text set in ITC Franklin Gothic Std.
Typeface provided by Adobe Systems.

Designer: Mary Ross
Lerner team: Sue Marquis

Library of Congress Cataloging-in-Publication Data

Names: Nickel, Scott, author.
Title: World's grossest creepy crawlers / Scott Nickel.
Description: Minneapolis, MN: Lerner Publications, 2022. | Series: World's grossest (updog books) | Includes bibliographical references and index. | Audience: Ages 8–11 | Audience: Grades K–1 | Summary: "Make your skin crawl with these disgusting creatures, from leeches to giant centipedes. Readers will come away with lots of knowledge about creepy critters, and they'll love grossing out their friends with facts and photos!"—Provided by publisher.
Identifiers: LCCN 2020010420 (print) | LCCN 2020010421 (ebook) | ISBN 9781728404004 (library binding) | ISBN 9781728418933 (ebook)
Subjects: LCSH: Insects—Juvenile literature.
Classification: LCC QL467.2 .N55 2021 (print) | LCC QL467.2 (ebook) | DDC 595.7—dc23

LC record available at https://lccn.loc.gov/2020010420
LC ebook record available at https://lccn.loc.gov/2020010421

Manufactured in the United States of America
1-48458-48972-6/16/2021

TABLE OF CONTENTS

ICKY INSECTS

These insects slither, squirm, and fly. They'll make your skin crawl.

Cockroaches gorge themselves on garbage. They leave behind piles of poop.

GORGE:
to eat large amounts

Dung beetles gather and eat poop.

Fire ants bite and leave you with pus-filled blisters.

BLISTER:
skin that is raised and has liquid inside of it

Flies throw up on their food and then slurp it up.

Scared stink bugs squirt smelly oil.

UP NEXT!

GRABBING A BITE.

BODY BUGS

Female mosquitoes feast on blood.

Head lice nest in hair and lay up to 150 eggs.

WHICH IS **GROSSER?**

Ticks vs. Fleas

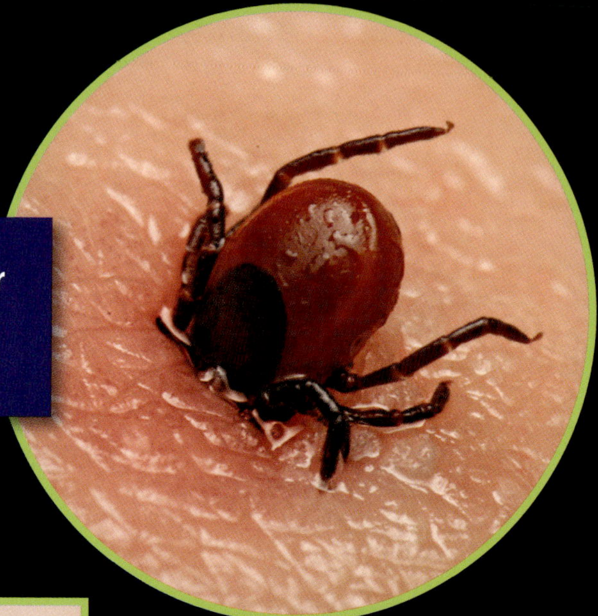

Ticks bury themselves under your skin. They suck blood until they are big and round.

Fleas also live on blood. They can jump fifty to one hundred times their body length.

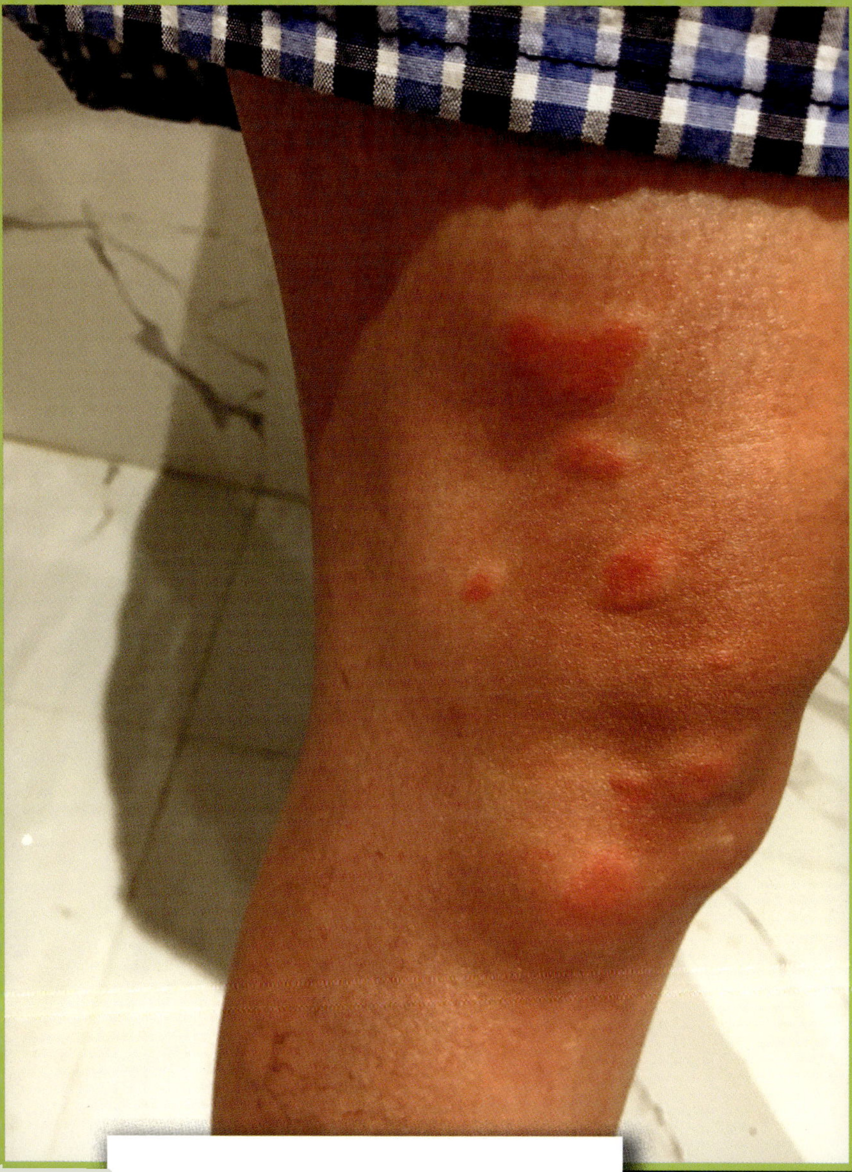

Mites burrow under the skin and can cause a terrible rash.

DISGUSTING DIAGRAM!

The assassin bug sucks all the blood from its prey. Some wear their dead prey's corpse.

wings
antennae
mouthparts
prey

Worms can eat dirt and poop. They can even live inside animals and humans.

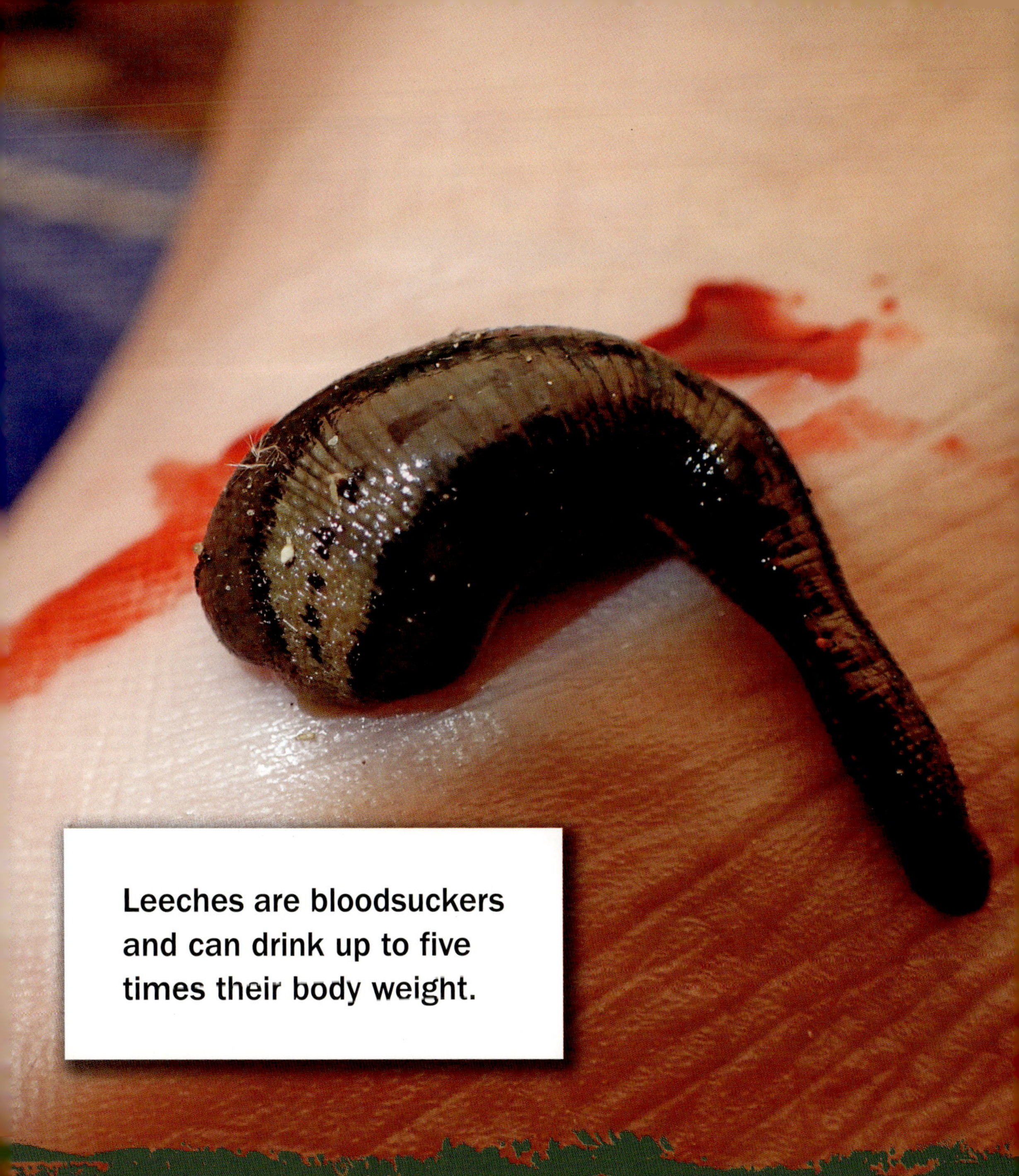

Leeches are bloodsuckers and can drink up to five times their body weight.

UP NEXT!

DANGEROUS AND DEADLY.

POISONOUS PESTS

Female black widow spiders sometimes kill and eat their male partners.

Tarantulas have hairy legs. They flick the hairs into a predator's skin.

PREDATOR: an animal that kills and eats other animals

Scorpions give birth to up to one hundred live babies.

The Amazonian giant centipede can grow up to 1 foot (0.3 m) long. These crawlers are disgusting.

AMAZONIAN: near the Amazon River in South America

Glossary

Amazonian: near the Amazon River in South America

blister: skin that is raised and has liquid inside of it

gorge: to eat large amounts

predator: an animal that kills and eats other animals

Check It Out!

Fiedler, Heidi. *Bugs: A Close-up Photographic Look Inside Your World.* Lake Forest, CA: Walter Foster Jr., 2017.

Markle, Sandra. *Mosquitoes: An Augmented Reality Experience.* Minneapolis: Lerner Publications, 2021.

National Geographic Kids: Insects
https://kids.nationalgeographic.com/animals/invertebrates/insects

Nickel, Scott. *World's Grossest Animals.* Minneapolis: Lerner Publications, 2022.

Pest World for Kids
https://pestworldforkids.org

What's That Bug?
https://www.whatsthatbug.com

Index

Photo Acknowledgments

Image credits: Design: Saravudh Sudkhaw/Shutterstock.com; Green angel/Shutterstock.com. Content: Ricky ku/Shutterstock.com, p. 4; IrinaK/Shutterstock.com, p. 5; Natalie colette kilian/Shutterstock.com, p. 6; SweetCrisis/Shutterstock.com, p. 7; Phil corley goldenorfephotography/Moment Open/Getty Images, p. 8; Photo Insecta/Shutterstock.com, p. 9; Pitiya Phinjongsakundit/Shutterstock.com, p. 10; khunkorn/Shutterstock.com, p. 11; Aksenova Natalya/Shutterstock.com, p. 12 (top); Mi St/Shutterstock.com, p. 12 (bottom); korn ratchaneekorn/Shutterstock.com, p. 13; Decha Thapanya/Shutterstock.com, p. 15; vitec/Shutterstock.com, p. 16; Kamol Kamtorncharoen/Shutterstock.com, p. 17; Sharon Keating/Shutterstock.com, p. 18; torook/Shutterstock.com, p. 19; I Wayan Sumatika/Shutterstock.com, p. 20; skifbook/Shutterstock.com, p. 21.

Cover: Jasius/Moment/Getty Images; Saravudh Sudkhaw/Shutterstock.com.